NUTRITION

FOR DOGS

MARYANN MOTT

Photo credits: Isabelle Francais, Alan Leschinski, Liz Palika, Robert Pearcy, Vince Serbin, Karen Taylor, Juanita Troyer, Wil de Veer

Distributed in the UNITED STATES to the Pet Trade by T.F.H. Publications, Inc., 1 TFH Plaza, Neptune City, NJ 07753; on the Internet at www.tfh.com; in CANADA by Rolf C. Hagen Inc., 3225 Sartelon St., Montreal, Quebec H4R 1E8; Pet Trade by H & L Pet Supplies Inc., 27 Kingston Crescent, Kitchener, Ontario N2B 2T6; in ENGLAND by T.F.H. Publications, PO Box 74, Havant PO9 5TT; in AUSTRALIA AND THE SOUTH PACIFIC by T.F.H. (Australia), Pty. Ltd., Box 149, Brookvale 2100 N.S.W., Australia; in NEW ZEALAND by Brooklands Aquarium Ltd., 5 McGiven Drive, New Plymouth, RD1 New Zealand; in SOUTH AFRICA by Rolf C. Hagen S.A. (PTY.) LTD., P.O. Box 201199, Durban North 4016, South Africa; in JAPAN by T.F.H. Publications. Published by T.F.H. Publications, Inc.
MANUFACTURED IN THE
UNITED STATES OF AMERICA
BY T.F.H. PUBLICATIONS, INC.

CONTENTS

NUTRITION AND YOUR
HEALTHY DOG
Page • 3

NUTRITION AND
HEALTH
Page • 41

NUTRITION FOR
DIFFERENT CANINE
STAGES
Page • 24

NUTRITION AND
HOUSEBREAKING AND
TRAINING
Page • 51

NUTRITION AND SKIN,
COAT, AND TEETH
Page • 35

CONCLUSION
Page • 60

SUGGESTED READING
Page • 64

NUTRITION AND YOUR HEALTHY DOG

Providing your dog with the right nutrition is one of the most important aspects of caring for your best friend. It's what helps him to fight off infection, maintain a shiny coat, and keep up his energy level. There are six essential nutrients that all dogs need in varying amounts in order to stay healthy.

PROTEIN

Protein helps serve numerous functions in the body including muscle growth, tissue repair, blood clotting, and immune functions. It is also burned as calories and can be converted to and stored as fat.

What dogs need is a subunit of protein called amino acids, of which there are two types: essential and nonessential. Nonessential amino acids can be made in the body, but essential amino acids must be supplied by food.

How much protein to feed on a daily basis depends on the quality of protein as well as a dog's life stage. In general, a healthy adult needs a minimum of 18 percent protein, on a dry matter basis. A dog that is still growing should have a minimum of 22 percent; lactating dogs, 28 percent; and high-performance dogs, 25 percent. If not enough protein is consumed, it can cause a rough dull coat, compromise the function of the immune system, contribute to poor growth, lower reproductive performance, and decrease milk production. In severe cases, weight loss and even death can occur. Because protein can be damaged by heat, it's important to buy food from a reputable manufacturer that uses proper cooking methods and employs quality control measures.

FAT

Fat supplies energy and is needed for the proper absorption of fat-soluble vitamins such as A, D, E, and K. It also supplies essential

A healthy dog is an active dog. Feeding your dog the proper nutrients is vital to his health and well-being.

fatty acids for a shiny, healthy coat and makes pet food tastier.

CARBOHYDRATES

Carbohydrates provide energy and keep the intestines functioning smoothly so that wastes efficiently pass through the body. Some common sources of carbohydrates found in pet foods are corn, oats, oatmeal, wheat, rice, and barley. A subunit of carbohydrates is fiber, which helps maintain proper function of the gastrointestinal tract. It is usually added in large amounts to specially formulated diets for weight loss because it gives a full feeling without adding extra calories.

Weigh-in time! A puppy should be weighed once a week to make sure that he is on the right track and growing properly.

VITAMINS

Vitamins are a natural substance that is only needed in small amounts. Fat-soluble vitamins (A, D, E, and K) need to be dissolved in fat to be utilized by the body. Water-soluble vitamins (Vitamin B group) need the presence of water for absorption and must be supplied daily to an active animal. So what do all these vitamins do? Here's a quick run down: Vitamin A plays a role in normal vision, reproduction, and formation of bones, and protects tissues from infection. Vitamin D is mainly used in the formation of bone tissue. Vitamin E is essential for a healthy reproductive system and healthy skin. The main function of Vitamin K is to help blood clot

These cute Akita puppies clamor around their bowl to gulp down every last bite.

As one of the six essential nutrients included in a dog's diet, carbohydrates help to give him the energy he needs.

properly. The ten vitamins that make up the B group are instrumental in the basic processes of respiration, digestion, and cell metabolism. And Vitamin C, which a dog's body can manufacture, helps with the production of collagen.

and producing hormones. Some pet food manufacturers will list an "ash content" on

the label, which tells you the total amount of minerals contained in the food.

Vitamins aid in normal vision, the reproduction of bones, and the fight against infection.

MINERALS

Generally, only small amounts of minerals are needed in pet food. The only exceptions are calcium and phosphorous to ensure proper bone formation. Minerals also help other bodily functions such as maintaining fluid balance and normal muscle and nerve function, transporting oxygen in the blood,

Though you may not realize it, water is an essential ingredient in a dog's diet. A big bowl of clean, fresh water should always be available to your dog.

WATER

You might not think of water as a nutrient but it is actually the most important one. Your pet can live without food for several days but just a 15 percent loss in body water can result in death. A big bowl of fresh, clean drinking water should always be available. Make sure to change it in the morning, then check it throughout the day. This is especially important for dogs that are subjected to hot weather when outside. Make sure to place the water dish in a shady area, otherwise the sun will make it too hot to drink. Also, if your backyard pooch likes to knock over his water dish, try filling a bucket with water then burying it halfway into the ground. This should keep it from being pushed over.

Now that you know what nutrients your dog needs, it's time to select a diet on which your dog can thrive. At first you might think it's a simple task. All you have to do is drive down to the local supermarket or pet store and pick out a brand, right? But once there you might be surprised at the dozens of choices you have. So how do you decide? Well, part of the question can be answered by the old adage, "You pay for what you get." People are usually attracted to generic foods because they're cheap. But in order to keep costs down, they use poorer quality ingredients. Generic brands

Just part of the scenery, these American Staffordshires are the picture of pure contentment.

Dive right in! This Australian Cattle Dog takes a break for a nice cool drink.

also contain fewer calories than premium brands; therefore, your dog must eat more of it to get the same amount of energy from a name-brand or premium food. Staying away from cheap brands doesn't mean you have to spend top dollar either. A product priced somewhere in the middle is usually your best bet. Owners also need to take into consideration a company's reputation. Manufacturers such as Hill's, Iams, Nature's Recipe, Purina, Waltham—just to name a few—have been in business for many years and regularly conduct nutritional research in their facilities and at top universities around the country. That knowledge goes back into improving the products.

When deciding what food to buy you also need to consider your dog's life stage: puppy, adult, nursing mother, working, or senior—these will be stated on the label. This is important because as your dog grows older his nutritional requirements change. Therefore, each life stage is formulated differently to meet these demands. For dogs with special needs (obesity) or that have a medical condition (heart, liver, or kidney) a specially formulated therapeutic diet recommended by a veterinarian will be needed.

By selecting a high-quality diet made by a reputable manufacturer that is for your dog's correct life stage, your choices on what prod-

uct to buy will be narrowed down considerably.

THE RIGHT CHOICE

But how do you know if you've chosen the right food for your dog? First, take a look at your dog's stool. You should see small, firmly formed feces with no fluid leaking out within 48 to 72 hours after starting the new diet.

If the food is not digestible, larger amounts of stool will be produced. You also don't want to see dry stool, which is less healthy for the large intestinal lining when trying to store and move feces. This can happen from high levels of fiber or minerals, or poor levels of protein.

Another sign that will tell you if a new diet is working

is a healthy, glossy coat, but this will take some time before showing up. A change in diet—good or bad—takes one to two months before it is reflected in the condition of your dog's coat.

CANNED, SEMI-MOIST, OR DRY

For busy people that work long hours, a dry formula is a good choice since it can be left in a bowl all day without spoiling. It's also the least expensive and some studies have shown eating dry kibble helps scrape tartar off teeth, which promotes good oral health. The downside is it's the least tastiest of the three.

Semi-moist food will not spoil at room temperature and comes in convenient prepackaged single servings so owners don't have to be bothered with measuring. The downside is the large amounts of sugar and preservatives added to maintain its freshness without refrigeration.

Canned food is by far the tastiest of the three but spoils after 30 minutes if not consumed. It is also the most expensive and requires more of it to be fed because the energy content is relatively low, especially for large breeds.

Some owners mix canned and dry food together. This common practice has prompted some manufacturers to include special feeding instructions for this on their labels. Ultimately, as long as your dog is receiving all the proper nutrients, the decision on which type to buy comes down to your preference, as well as his.

SETTING STANDARDS

There are two agencies that work together in regulating pet food labels. The Food and Drug Administration's Center for Veterinary Medicine establishes and enforces standards for all animal feeds. This federal agency oversees such labeling aspects as proper identification of product, net quantity statement, and listing of ingredients.

The second agency is the Association of American Feed Control Officials (AAFCO), which is a non-government group made up of state and federal officials from around the country. They establish pet food regulations that are more specific and cover such

A shiny coat, a healthy appetite, and an energetic approach to life are all signs that your dog is happy and in good physical and mental condition.

There are many kinds of dog food to choose from—canned, semi-moist, or dry. For these hungry Akitas, a combination of food works best.

things as the guaranteed analysis, the nutritional adequacy statement, and feeding directions. Each state decides whether or not to institute or enforce AAFCO's regulations. Most do, but some don't.

READING LABELS

When selecting the right food to buy for your best friend, learning how to decipher what labels say (or don't say) is important. Especially when you consider there are dozens of different brands to choose from every time you step inside a grocery or pet supply store.

Product Name

What's in a name? When it comes to pet food, a lot. Specific words used in a product's name can surprisingly reveal what is actually in the food or not. For example, if a product is called "Tom's Beef Dog Food," it must contain 95 percent beef. But if it is called "Tom's Beef *Dinner* for Dogs" it is only required to contain a minimum of 25 percent beef. That's a big difference for adding just one word. Besides dinner, other terms such as platter, entree, nuggets, and formula also fall under this 25 percent minimum requirement. When these words appear, check the ingredient list. You might be surprised to find out your so-called beef dinner is really made up of mostly chicken. Also, if more than one ingredient is included in a dinner name like "Beef and Chicken Dinner for Dogs," they both must total 25 percent and be listed in the same order as found on the ingredient list.

Another tricky word to watch for is "with." If a product is called "Tom's Dog Food *With* Beef," it only has to contain 3 percent beef. Originally, the term "with" was intended to highlight minor ingredients outside of a product's name, for example a side burst saying, "with bacon." But recent amendments to AAFCO's regulations now allow the use of "with" in a product's name.

Unlike the other terms above, when "flavor" is used in a product's name there is not a minimum percentage required. The product only has to contain a sufficient amount of the flavor to be detectable. In order to confirm this, test methods using animals trained to like a particular flavor are used. When you see a product called "Beef Flavored Dog Food," it might contain beef but more than likely it is another substance giving it the flavor, like beef meal or beef by-products.

As a pet owner, it is your responsibility to know the ingredients that go into the food you give your dog.

Ingredient List

Unfortunately, the quality of ingredients are not listed here, only the names. Each ingredient is listed in descending order according to weight. Look for animal-based proteins high up on the list such as chicken or chicken by-products, lamb, lamb meal, fish meal, and egg. But since manufacturers know you do this, read the label carefully. Some manipulate certain ingredients in order to place it lower on the list and a more desirable one, such as protein, first. This is usually done by dividing a grain ingredient into different categories, such as wheat flour and whole ground wheat, or ground rice and rice bran. Splitting the grain into different categories reduces its weight and pushes it lower on the list.

Owners should also look on the list for large amounts of sugar or salt since some manufacturers add them to make the food more tasty.

Another ingredient to watch for is ethoxyquin, which has been a controversial food preservative used over the last ten years. The Center of Veterinary Medicine (CVM) has received some complaints from dog owners that ethoxyquin has caused a variety of problems such as allergic reactions, skin problems, major organ failure, behavior problems, and even cancer. The CVM found very little scientific information to support these claims and will not lower the maximum allowable levels until more evidence is available. Currently, more studies are being conducted. In the meantime, though, CVM has asked the pet food industry to voluntarily lower the maximum level of ethoxyquin in dog foods from 150 ppm to 75 ppm.

When reading an ingredient list you'll probably come across some of the common ones listed below. And even though some of these don't sound tasty to us, they still provide essential nutrients for our pets.

Meat is the clean flesh derived from slaughtered mammals and is limited to that part of the striate muscle which is skeletal or that is found in the tongue, in the diaphragm, in the heart, or in the esophagus.

This adorable Lab is the picture of what a healthy pup should look like.

Life is pretty good for this healthy and happy Australian Cattle puppy.

A responsible pet owner should be knowledgeable in the area of dog food. Meat, meat by-products, and poultry by-products are just some of the terms you should be familiar with.

Feeding Instructions

Keep in mind the amounts recommended on the label are only suggestions. Some dogs will need more; some less. Breed, age, and activity level all play a role in how much food your dog will need. Start off with the amount suggested on the label but don't be afraid to increase or decrease when necessary. Also keep in mind the amount stated is for the entire day. To figure out the correct serving size for a single feeding, divide the suggested amount by how many times a day (two or three) you feed your pet.

Calorie Statement

New regulations now allow manufacturers, on a voluntary basis, to include a calorie statement on labels. Calories are expressed on a "kilocalories per kilogram" basis.

Meat by-products are the nonrendered lean parts, other than meat, derived from slaughtered mammals. It includes, but is not limited to the lungs, spleen, kidney, brain, liver, blood, bone, stomach, and intestines freed of their contents.

Poultry by-product meal consists of the ground, rendered clean parts of the carcass of slaughtered poultry, such as necks, feet, undeveloped eggs, and intestines.

Meat and bone meal is the rendered product from mammal tissues, including bone, exclusive of any added blood, hair hood, horn, hide trimmings, manure, stomach, and rumen contents, except in such amounts as may occur unavoidably in good processing practices.

Not only is the type of food you feed your dog important, but the amount is as well. Portions of the food will vary depending on the breed of dog, his activity level, and age.

Kilocalories are the same as calories, and kilogram is a metric measurement that equals 2.2 pounds. If a calorie statement is not included on the label, it can be figured out by doing the calculation in the chart. (To find the percentages you need for this calculation, look on the label under the guaranteed analysis section.)

Step 1: Multiply the percentage of crude protein by 3.5 then write down the answer.

Step 2: Multiply the percentage of crude fat by 8.5 then write down the answer.

Step 3: Add the percentages of crude protein, crude fat, crude fiber, moisture, and ash. Then subtract the total from 100. This gives you the percent of nitrogen-free extract (NFE), which is the carbohydrate portion.

Step 4: Multiply the NFE percent by 3.5 and write down the answer.

Step 5: Add the answers from steps 1, 2, and 4. Then multiply the total by 10. The answer is the amount of calories.

Guaranteed Analysis

This shows the minimum percentage of crude protein and crude fat as well as a maximum percentage of moisture (water) and crude fiber. The term "crude" refers to the method of testing the product, not to the quality of the nutrient. Manufacturers sometimes list other nutrients such as ash or calcium, though these are not required. Keep in mind only the minimum and maximum amounts are stated; therefore, actual amounts can vary widely.

Example:	
Crude protein	24% x 3.5 = 84
Crude fat	10% x 8.5 = 85
Crude fiber	3%
Moisture	10%
Ash	5%
Total	**52%**

NFE (100-52 = 48)
48% x 3.5 = 168
84 + 85 + 168 = 337
337 x 10 = 3370 kcal/kg

When comparing the guaranteed analysis percentages between a canned and a dry product, the canned food will usually show lower amounts of protein and other nutrients. This is because of the difference in moisture. Canned foods contain up to 78 percent water while dry foods only contain 10 to 12 percent. To compare nutrients between foods with different moisture levels, some math work is required. Both products must be converted to a dry-matter basis, which is the portion of food left if all the water were taken out.

Dry-Matter Math

This example shows how to convert the percentage of protein found in a canned

This curious Shar-Pei sets his sights on what he wants and goes after it.

product to a dry-matter basis. First subtract the percentage of moisture shown under the label's guaranteed analysis section from 100. Next, take the percentage of crude protein and divide it by the percentage just calculated; then multiply the answer by 100. This gives you the percentage of protein on a dry-matter basis. Now in order to make your comparison you'll have to perform the calculation again for the dry food.

Example: 100 - 75 = 25 percent dry-matter basis
8/25 = .32
.32 x 100 = 32 percent protein on a dry-matter basis.

Nutritional Adequacy Statement

This tells you what life stage the product is formulated for such as growth, reproduction, maintenance, senior, or all life stages. It should also state whether the food is "complete and balanced" or "complementary." Complete and balanced means it contains all the essential ingredients your dog needs on a daily basis to stay healthy and can be fed by itself as a meal. Complementary means it is not intended to be used as a meal and must be mixed in with other food to create one.

According to AAFCO guidelines, there are three ways a manufacturer can claim their product is complete and balanced, which are listed below. If one of the following statements is not found under the nutritional adequacy section of the label, don't buy the product.

The first says, "Animal feeding tests using Associa-

tion of American Feed Control Official procedures substantiate that (XYZ product) provides complete and balanced nutrition for (XYZ life stage)." This means the food was fed to actual dogs during the life stage claimed and found to be adequate.

The second way states "(XYZ product) is formulated to meet the nutritional levels established by the AAFCO Food Nutrient Profiles for (XYZ life stage)." This means the food was analyzed in a laboratory and compared to a recognized industry standard.

The third way, which is a new AAFCO statement, says "(XYZ product) provides complete and balanced nutrition for (XYZ life stage) and is comparable in nutritional adequacy to a product that has been substantiated using AAFCO feeding tests." This means the manufacturer has not tested the product, only a similar one that was found to be complete and balanced. AAFCO has decided to allow this because some manufacturers produce food that is made exactly the same way, the only difference is the flavor—beef, chicken, or lamb. This statement allows manufacturers to avoid the high cost of testing each individual flavor in order to put a complete and balanced claim on the label.

Manufacturer Information

This section of the label identifies the company responsible for the quality and safety of the product. Most manufacturers list a toll-free number or address here for consumers' questions or complaints. Also, some will list web sites that contain vast

A good way to ensure that new puppies will be born healthy is to take excellent care of the mother.

amounts of information about their products.

Net Quantity Statement

This shows the weight of food contained in a bag or can in both pounds or ounces and in a metric measurement such as kilograms or grams. This information is especially important when buying dry food since some companies use bags that can hold 40 pounds of food, but then only put 35 pounds in it.

WHAT LABELS DON'T TELL YOU

Unfortunately, not everything we need to know is stated on the label. For instance, it doesn't say if the food is made from a variable or fixed formula. A variable formula is where nutrients are consistent from batch to batch, but ingredients vary. A fixed formula uses the same ingredients in every batch.

Some wording used on labels can also be misleading. Food touted as gourmet or premium, for example, is not required to contain any higher quality ingredients or nutritional standards than any other complete and balanced product. Another misleading word is "natural." Some people might think this means minimally processed, but according to a veterinarian at the FDA, not all pet foods meet this criteria. Some consumers might also think natural means the product contains no artificial ingredients, but all complete and balanced products must contain some chemically synthesized ingredients such as vitamin supplements.

FOOD STORAGE

Rancidity is caused by four main factors: heat, humidity, light, and oxygen. If eaten, spoiled food can cause vomiting, diarrhea, and other serious health problems. That's why it's important for owners to learn how to store food properly.

Dry food—opened or unopened—should be kept in a cool, dry place. Make sure to tightly seal any unused portions. You can find specially designed food storage containers with airtight lids at some pet supply stores. When

Although it's difficult to see through all that fluff, this Chow looks like he is properly nourished.

properly kept, dry food is guaranteed fresh up to 12 months from the date of manufacture.

Unopened canned products can last 18 months from the date of manufacture. But keep in mind, once opened it should be immediately refrigerated and used within two days.

Besides proper storage, owners should look for a "best if used by" date on the label to ensure the product is fresh. More manufacturers are now adding this important piece of information on their products. Also, when buying food, shop at stores that have a high turnover in their pet food

inventory. This way if the product doesn't have an expiration date, you won't be stuck with one that has been sitting on the shelf too long.

CHANGING FOOD

When switching brands it's best to take a gradual approach to reduce or eliminate the possibility your pet will suffer an upset stomach or diarrhea. To do this, slowly combine increasing amounts of the new product with decreasing proportions of the old until only the new product is fed. This should be done over a seven to ten day period. While doing this, make sure the amount fed does not change, only the proportions of the new product versus the old.

TABLE SCRAPS

It's okay to give your dog healthy table scraps now and then—just don't over do it. Snacks should only comprise 10 percent or less of your dog's daily caloric intake. By feeding too many treats throughout the day it can create a finicky eater or, worse yet, cause your dog to pack on extra pounds. To avoid this, subtract from his daily meal what you add in the form of treats. Some snacks you and your pooch can share are carrots, broc-coli, cauliflower, cooked zucchini, apple slices, raisins, and seedless grapes. Pop-corn—without butter—and ice cubes, the ultimate low calorie treat, can also be given. Other healthy choices are cooked fish, chicken, or lean beef, but avoid chicken skin or lunch meats that are high in fat and calories. Foods to avoid are hot dogs, ice

It's important for pet owners to learn how to store food. Spoiled food can cause vomiting, diarrhea, and other serious health problems.

cream, cookies, cake, and above all else—chocolate, which can be extremely toxic.

BONES

Some dogs just love to chew on bones. In fact, experts say it can even help scrape tarter off teeth, which in turn helps promote good oral health. But certain bones should be avoided, like turkey, pork, or chicken, which can easily splinter causing digestive upsets and intestinal blockage. If you want to give him a bone, some good choices are knuckle or marrowbones, as long as they are partially boiled first. But remember if at any time your pet starts to eat the bone rather than just chew on it, take it away.

GRASS

Important medical and nutritional advances have been made over recent years, but when it comes to something as simple as why dogs love to graze on grass, experts don't have an answer. One theory is dogs are lacking an essential nutrient in their diets, but grass doesn't contain any useable nutrients. Others say it's because a dog has an upset stomach and needs to vomit. But perfectly healthy dogs munch on grass too, not just sick ones. The most logical explanation is they just simply like the taste. Whatever the reason, it's okay to let your dog eat grass as long as you're careful of pesticides and fertilizers that, if ingested, can make your dog extremely sick. If you live in an area where the grounds are maintained for you, check to see what kind of chemicals are applied and when. Even if

Bad habits are hard to break! Although it's acceptable to give your dog table scraps occasionally, making it a habit is not wise.

your dog doesn't nibble on the grass, just walking across a freshly treated section can be harmful if he licks his paws afterwards.

CAT FOOD

Why do dogs eat cat food? Manufacturers know cats are finicky eaters, so in order to tempt their taste buds, they formulate food with a titillat-

ing smell. One whiff of the strong aroma wafting through the air and dogs can't resist.

HOMEMADE DIETS

The field seems to be evenly divided on this somewhat controversial issue. Some experts say home-cooked meals do not provide all the necessary or proper amounts of nutrients your pet needs to

Not only do dogs love to chew on bones, experts say it can even help scrape tartar off teeth, which promotes good oral health.

extent, both sides are right. If you have the extra time, the money, and the belief that it is healthier for your pet to eat a home-cooked meal, then you should do it. Just make sure you go to a reputable source for a nutritionally balanced recipe. Your veterinarian might be able to suggest a recipe or at least give you information on where to find one. For owners whose lives are extremely busy, you're best sticking with a high-quality commercial diet. Each year pet food manufacturers spend millions on nutritional research and despite their critics, many dogs lead happy, healthy lives on a commercial formula.

VEGETARIAN DIETS

Another debate also rages on over whether a dog can live a healthy life on a vegetarian (meatless) or vegan (animal product-free) diet. On one side of the debate is the mainstream veterinarian community that says dogs are carnivores and therefore need to eat meat. They contend that if meat is excluded from a dog's diet it can have a negative effect on his general health.

On the other side, animal rights activists and holistic veterinarians say dogs don't experience any health problems. And besides, the meat found in dog food is deemed unfit for human consumption, they say, so if we wouldn't eat it, why should our pets?

For vegetarians the decision is a tough one. They're torn between keeping their commitment to a humane lifestyle yet wanting to do

stay healthy. And, they add, shopping for ingredients and then having to cook it is more of a hassle and expense than simply ripping open a bag of kibble.

But those in favor of homemade diets say it's worth the extra aggravation. They claim commercial pet foods are filled with moldy grains, rancid animal fat, and contaminated meat. These questionable ingredients are just downright unhealthy for our pets and the only way around it is for owners to take the matter into their own hands and cook for their pets. To an what is best for the health and well-being of their dog.

Recently the People for the Ethical Treatment of Animals conducted a survey to settle the question and quell the nagging doubts for those who choose to provide their dogs with a vegetarian diet. During a one-year period they gathered and analyzed data on the diet and health status of 300 vegetarian dogs. From this study they concluded the following:

—The longer a dog remains on a vegetarian or vegan diet, the greater the likelihood of overall good to excellent health. It's also less likely he will get infections, cancer, or hypo–thyroidism.

—A vegetarian diet may increase the alkalinity of a dog's urine, promoting susceptibility to urinary tract infections, which can be prevented by using cranberry capsules.

—The longer a dog remains on a vegetarian or

Although there are many well-respected brands of dog food to choose from, some pet owners still opt for homemade diets. This Golden Retriever is more than happy to oblige.

Why dogs love grass so much is still a mystery to experts, but it's not harmful to your dog as long as you're careful of pesticides and fertilizers.

Stainless steel bowls are inexpensive, lightweight, unbreakable, and easy to keep clean. These English Foxhound pups don't seem to mind the choice.

vegan diet without supplementation of L-carnitine or taurine, the greater the likelihood that he could come down with dilated cardio–myopathy.

—Dogs without soy foods in their diet appear to be in better health than those who eat soy.

If you can't decide what to do, one expert recommends a compromise of sorts. He says to use more poultry, eggs and dairy products than beef, since their production consumes fewer natural resources. And if possible, turkey should be selected over chicken because it is usually raised more humanely.

FEEDING SUPPLIES

Decisions, decisions, decisions. When it comes to

There are many varieties of diets that you can choose to give your canine companion; however, it has to agree with your dog and his lifestyle.

Keeping your dog food in a dry, clean container with a lid is a good choice, otherwise your hungry pup might just help himself.

buying a bowl for your dog there are plenty of choices. Some ceramic bowls come in fancy colors and designs and besides looking good, these heavy dishes make it tough to shove around the floor or knock over while eating. There are also stainless steel or plastic bowls that are inexpensive, lightweight, and unbreakable. Some even come with rubber stripping around the bottom to keep it in place, while others are designed to prevent floppy ears from getting wet or smeared with food.

If you have a large dog, consider purchasing an elevated food stand, which will be more comfortable for your dog to eat from. Some models can even be adjusted to different height levels making it easy for owners with large-breed puppies to modify it as they grow. Raised eating dishes are also recommended for dogs prone to bloat, or older dogs that are afflicted with arthritis and experience pain when bending down low to reach their food.

"Just a little more, I know I can get it!" This Gordon Setter has set his sights on those yummy pretzel sticks.

Another specially designed dish has a moat around it that can be filled with water to prevent ants and other crawling insects from getting into the food. For dogs that stay outside in colder climates there are heated bowls to prevent his drinking water from freezing.

Whichever you choose, you'll need to buy one dish for food and one for water. If you have more than one pooch, you'll need to supply each with their own bowls. This helps prevent fights from breaking out and lets each dog get the proper amount of food to eat.

Remember after every meal to wash the bowls with warm, soapy water or run them through the dishwasher. This will prevent bugs and bacteria build-up.

Here's a Tip!

Owners should be careful when using products attached to outside water faucets that dogs can lick for a drink. After a couple of hours in the sun, the metal nozzle can become extremely hot and burn your dog's tongue. Worse yet, if this is his only source of water all day, your dog can become dehydrated.

Although these puppies are sharing their bowl nicely, it's a good idea to purchase separate bowls for each dog to prevent unnecessary fighting.

Be certain that your dog's nutritional requirements are properly met. Taking note of the National Adequacy Statement will tell you what life stage the product is formulated for and if the food is "complete and balanced."

NUTRITION FOR DIFFERENT CANINE STAGES

As your dog goes through different stages in life, his nutritional needs change. Unfortunately, there are no hard and fast rules that state exactly when an owner should adjust their pet's diet to meet these changes, but by having yearly check-ups, your veterinarian can help advise you when to make a switch.

PUPPY STAGE

It's nonstop action at this stage—not only from all the playing that puppies love to do but also from the rapid growth that takes place. That's why larger quantities of nutrients like protein and fat are needed for extra energy.

Owning a puppy is a big responsibility. Before you purchase your puppy, make sure you have enough time and patience to devote to the newest member of your family.

There's nothing more adorable than a trio of puppies—like these fuzzy Akita babies!

When selecting a product to feed your puppy, look for one that says it is complete and balanced for growth. By doing this, supplements are not needed and, in fact, may even be harmful. If you are considering adding supplements, consult your veterinarian first.

If your puppy came from a breeder, find out what kind of food, the quantity, and the time he was fed. Because they are creatures of habit and feel more comfortable with familiar routines, try to stick with this schedule until your puppy gets adjusted to his new home. Also keep in mind that a change in diet can cause an upset stomach and diarrhea. If you have to change brands, gradually mix increasing amounts of the new food with decreasing proportions of the old food over a period of several days.

If you don't know the prior feeding schedule, you'll need to first figure out how much to feed. This will depend on his size, activity level, and breed. Start off by following the feeding directions on the label and then increasing or decreasing as needed. Remember the amounts stated are for the entire day. For a single feeding, divide the suggested amount by how many times a day (two or three) you feed your pet.

Because puppies have little stomachs, they should be fed several times a day in small proportions. Establish a routine by serving meals at the same times and place each day. Up until four months of age, puppies will need four meals per day: morning, midday, mid-afternoon, and an hour before bedtime. At four to six months of age, reduce to three meals a day. From eight months on, feed once or twice a day depending on your schedule and what best suits your pet. For sanitary reasons, wash the bowl with warm soapy water after every meal. And don't forget to provide plenty of fresh, clean drinking water all day.

Proper feeding habits are extremely important during this stage, especially for large- (more than 65 pounds when mature) and giant-breed puppies (more than 90 pounds when mature) that can suffer from skeletal abnormalities like hip dysplasia, hypertrophic osteo–dystrophy, and osteo–chondrosis if they grow too fast.

Just like children, puppies are both curious and playful. They need constant attention and a watchful eye to make sure they don't get themselves into harmful situations.

These diseases usually become apparent during the rapid growth phase, which is between four to six month of age.

Hip dysplasia is a hereditary disease caused by the improper development of the hip joint. It strikes any breed but tends to show up more in larger ones. Chronically affected dogs may only experience a mild discomfort, especially after intense exercise. Those with severe hip dysplasia have an abnormal gait, experience pain and lameness, and have difficulty rising.

Hypertrophic osteo–dystrophy predominately affects breeds like the Great Dane, Irish Wolfhound, Saint Bernard, Dalmatian, Weimaraner, Doberman Pinscher, Labrador Retriever, Collie, and Greyhound. Experts say it commonly shows up around three to six months of age and is characterized by anorexia and lameness.

Osteochondrosis dissecans is a genetic disease prevalent in large and giant breeds like the Great Dane, Labrador Retriever, Newfoundland, and Rottweiler. Experts say it is characterized by degeneration of bone underlying the articular cartilage of joint surfaces and can cause lameness.

To an extent all three of these problems can be nutritionally managed by special diets designed to control the rate of growth in large-breed and giant-breed puppies. It

should be noted, however, this food does not affect how big the puppy will ultimately become once mature—genetics have already determined that—it only slows down their rate of growth to help prevent these skeletal problems. Research shows a diet containing a minimum of 26 percent protein from a high-quality, animal-based source and a minimum of 14 percent fat is best. Some experts say large-breed and giant-breed puppies should also consume 25 percent less calcium than smaller breeds, since too much calcium can also increase the dog's risk of developing musculoskeletal problems.

For puppies that are small, large, or in-between, improper

feeding also increases the risk of obesity, contributes to poor muscle and bone development, and weakens the immune system. To make sure a puppy is on the right track and growing properly, he should be weighed once a week. His growth rate should then be compared with published charts for that breed, but getting your pet to step on a scale might be difficult. Here's a tip: weigh yourself first, then weigh yourself again while holding the puppy. The difference between the two numbers is your puppy's weight.

Eventually when your puppy nears the size and weight of an adult dog, which can be anywhere from six months to two years

At the senior stage in a dog's life, his metabolism and activity level begin to slow down. But like puppies, aging dogs still need a lot of love and affection.

Together, we can do no wrong! Laurence and his Staffordshire Bull Terrier keep an eye on each other.

depending on breed, gradually introduce an adult formulated food.

ADULT DOG

For a healthy adult dog, feeding requirements are straightforward and simple. Experts say a diet should contain, on a dry-matter basis, 18 percent protein and 5 percent fat. How much to feed depends on size, activity level, and environment. Start off by following the directions on the label, then increasing or decreasing as needed. Dogs eight months of age and older can be fed once or twice a day depending on your schedule and what best suits your pet. But if you have a working dog (sheep, police, field) or one that is convalescing, more than one meal is needed. A consistent feeding schedule should be established by serving meals at the same time each day and providing plenty of fresh drinking water. If your dog's appetite varies from day to day, don't worry—this is normal. But if it doesn't pick up again soon or he shows signs of illness, consult your veterinarian.

SENIOR DOG

At this stage a dog's metabolism and activity level start to slow down and a change in diet is necessary. A senior dog's diet should have fewer calories and include fatty acids, as well as antioxidants like beta carotene and Vitamin E to help support the immune system. In the past, some experts thought older dogs needed less protein, but that has proven not to be true. Protein levels need to remain the same for a healthy adult dog in order to maintain muscle condition. Knowing when to switch to a senior diet depends on your dog's size, because larger breeds age faster than smaller ones. The list below shows the age when certain size dogs are considered "old":

Small breeds under 20 pounds (Yorkshire Terrier, Chihuahua)—8 to 10 years of age.

Medium breeds 20 to 50 pounds (Airedale Terrier, Cocker Spaniel)—7 to 9 years of age.

Large breeds 50 to 80 pounds (German Shepherd, Doberman)—6 to 8 years of age.

Giant breeds over 80 pounds (Great Dane, Saint Bernard)—5 to 7 years of age.

As dogs age and become less active, some tend to become overweight. Those extra pounds put more strain on the heart, lungs, muscles, and joints. In fact, obese dogs have an increased risk of kidney and heart disease and may even die earlier. If your dog has packed on a few too many pounds, consult your veterinarian about starting him on a low calorie diet and exercise program.

On the other end of the scale, some dogs become too thin due to digestive problems that prevent the body from utilizing all the nutrients in their diet. One way to help beef up these skinny seniors is to give small, frequent meals instead of serving just one large meal.

Tip: Dealing with Old Age

If arthritis starts making it difficult for your dog to bend down to reach his food bowl, try using an elevated food stand, or place the bowl on a chair or step that is a comfortable height.

PREGNANCY

During the entire pregnancy, owners should feed a good-quality diet that states on the label that it is for reproduction and growth. Some experts suggest a diet that contains 8 percent fat for extra energy and 22 percent protein (on a dry-matter basis), which is important for milk production. From the fifth week on, increase the amount of food given daily by 10 to 15 percent. By the time she gives birth (whelps) she should be eating 35 to 50 percent more than her normal ration.

After the puppies are born, allow food to be available all day. It is important that nursing females eat all they want during this time so they can produce enough milk for their puppies. At peak lactation (milk production) up to 7 percent of her body weight, per day, is given to her puppies in the form of milk. If the mother refuses to eat dry food, try serving canned, which is more palatable, several times throughout the day—but remember to remove it if not consumed after 30 minutes since it spoils.

Also make sure to provide plenty of fresh clean drinking water, which aids in milk production. Encourage consumption by always keeping bowls clean and changing water frequently throughout the day. Remember, if you wouldn't drink it, why would your pet?

The demand for milk by nursing puppies will continue to increase for up to four weeks. Normally, by six to

This mommy Chow and her pup are sitting pretty in the afternoon sun.

While a female is nursing, it's important that she is allowed to eat all that she wants so she can produce enough milk for her puppies.

eight weeks of age, puppies are weaned and will start to eat solid foods. At this point, gradually lessen the amount of food given to the mother until you reach the normal ration she received before becoming pregnant.

BREED GROUPS

Besides nutritionally managing each life stage, one pet food manufacturer has recently formulated special diets for each breed group: Sporting, Hound, Working, Terrier, Toy, Non-Sporting, and Herding.

These newly introduced diets can help prevent and manage certain disorders common to the breeds in each group. And even though each

of the seven diets contain different ingredients, they all still share the following:

Parsley, Spearmint and Ginger—settles the stomach.

Gingko Biloba—improves blood and oxygen flow, enhances circulation, regulates blood glucose and blood pressure.

Evening Primrose, Salmon and Canola Oils—all three are combined to manage inflammation associated with allergic reactions and promote good coat and skin condition.

Chicory Root—an extract from endive, it helps promote gastrointestinal health.

Eyebright—increases blood circulation in the eyes and, as its name suggests, keeps eyes clear.

Yucca Schidigera Extract— aids elimination and helps reduce stool odor.

Brewers Yeast (Chromium)—combined with niacin, manages blood sugar levels and helps to prevent hyperglycemia.

Sporting Group

This group includes the following breeds: the Brittany, Pointer, German Shorthaired Pointer, German Wirehaired Pointer, Chesapeake Bay Retriever, Curly-Coated Retriever, Flat-Coated Retriever, Golden Retriever, Labrador Retriever, English Setter, Gordon Setter, Irish Setter, American Water Spaniel, Clumber Spaniel, Cocker

Dogs in the Working Group, like this Doberman Pinscher, commonly encounter heart, gastrointestinal, and bone and joint difficulties. To nutritionally manage these problems, hawthorn berries, taurine, and selenium are added to their diets.

Spaniel, English Cocker Spaniel, English Springer Spaniel, Field Spaniel, Irish Water Spaniel, Sussex Spaniel, Welsh Springer Spaniel, Vizsla, Weimaraner, and Wirehaired Pointing Griffon.

This group usually needs extra calories and energy for exercise performance. To provide this, higher amounts of vitamin and mineral fortification as well as choline are added to help process nutrients into energy. Some Irish Setters are gluten intolerant, which can cause weight loss and chronic diarrhea. To combat this, gluten containing grains like wheat, barley, and oats are eliminated from the diet and only rice is used, which makes the product gluten-free.

Another condition found in the Sporting Group that affects Cocker Spaniels and some Golden Retrievers is a low blood taurine level that has shown to be responsive to supplementation. Several other ingredients in the product also help and act as antioxidant sources like Vitamin E, selenium, rosemary, and hawthorn berry powder, which has been reported by some supplement manufacturers to promote heart health.

Several breeds in the Sporting Group also tend to have hip dysplasia. Studies have shown that slowing down the growth rate of these breeds while puppies has a positive impact in reducing this hereditary developmental disease. It is recommended that you reduce the feeding requirements for those prone by 15 percent. This reduction is also encouraged for breeds like Cocker Spaniels and Labradors to reduce weight gain during growth and help

prevent obesity as they get older.

Working Group

This group includes the following breeds: the Akita, Alaskan Malamute, Bernese Mountain Dog, Boxer, Bullmastiff, Doberman Pinscher, Giant Schnauzer, Great Dane, Great Pyrenees, Greater Swiss Mountain Dog, Komondor, Kuvasz, Mastiff, Newfoundland, Portuguese Water Dog, Rottweiler, St. Bernard, Samoyed, Siberian Husky, and Standard Schnauzer.

This group commonly encounters heart, gastrointestinal, and bone and joint difficulties. To nutritionally manage these problems, hawthorn berries, taurine, and selenium are added to the product. For heart health, the formula is restricted in salt. Ground rice, rolled oats, and cracked pearled barley is added to help with digestibility. Glucosamine is included to promote bone and joint

The energetic Dalmatian is part of the Non-Sporting Group. Food for Dalmatians often contains alkin urine because the dogs tend to have uric acid calculi problems.

As part of the Herding Group, your German Shepherd will need food that can carefully manage his fat calories.

(Miniature and Standard), Schipperke, Shiba Inu, Tibetan Spaniel, and Tibetan Terrier.

This formula contains alkin urine for Dalmatians because they tend to have uric acid calculi problems. Also, since Poodles and Bichons tend to have more problems with periodontal disease, sodium hexametaphosphte is added, which has been shown in clinical studies to reduce tartar accumulation by approximately 50 percent. A generous supply of Vitamins A and B and minerals such as copper and zinc are added to promote good skin and hair coat condition for breeds in this group.

Herding Group

This group includes the following breeds: Australian Cattle Dog, Australian Shepherd, Bearded Collie, Belgian Malinois, Belgian Sheepdog, Belgian Tervuren, Border

health, while chicory root aids digestion.

Non-Sporting Group

The following breeds are included in this group: American Eskimo Dog, Bichon Frise, Boston Terrier, Bulldog, Chinese Shar-Pei, Chow Chow, Dalmatian, Finnish Spitz, French Bulldog, Keeshond, Lhasa Apso, Poodle

The American Eskimo is a playful member of the Non-Sporting Group.

Collie, Bouvier des Flandres, Briard, Caanan Dog, Collie, German Shepherd Dog, Old English Sheepdog, Puli, Shetland Sheepdog, Cardigan Welsh Corgi, and Pembroke Welsh Corgi.

German Shepherds tend to suffer chronic intermittent diarrhea, which comes from an intestinal immune deficiency. Therefore, in this formula, the level of fat calories is carefully managed. This is important since fat digestion is the first to go if the intestinal track starts to fail. Two ingredients also added to this diet are bromelain, which is a pineapple extract, and ginger to help with digestion. Also included are glutamine, which encourages small intestine repair, and chicory root to maintain gastrointestinal health.

Hip dysplasia is prevalent in several breeds of the Herding Group like Collies and Shetland Sheepdogs. Studies have shown that slowing down the growth rate of these breeds while puppies can reduce incidents of this bone problem. A 15 percent reduction from the standard feeding directions is recommended. This reduction also helps breeds that are prone to obesity, like Old English Sheepdogs, stay slim.

Terrier Group

This group includes: the Airedale Terrier, American Staffordshire Terrier, Australian Terrier, Bedlington Terrier, Border Terrier, Bull Terrier, Cairn Terrier, Dandie Dinmont Terrier, Smooth and Wire Fox Terrier, Irish Terrier, Kerry Blue Terrier, Lakeland Terrier, Standard Manchester Terrier, Miniature Bull Terrier, Miniature Schnauzer, Norfolk Terrier, Norwich Terrier,

English Bull Terriers are part of the Terrier Group.

Scottish Terrier, Sealyham Terrier, Skye Terrier, Soft Coated Wheaten Terrier, Staffordshire Bull Terrier, Welsh Terrier, and West Highland Terrier.

In this group, 75 percent of all Bedlington Terriers have copper storage disease or are carriers of it. Therefore, copper levels are managed in this formula. Also, high levels of zinc are added to encourage copper excretion from the body. Milk Thistle, which is used by Europeans suffering from cirrhosis of the liver, is also included to promote liver repair.

Toy Group

This group includes the

following: the Affenpinscher, Brussels Griffon, Chihuahua, Chinese Crested, English Toy Spaniel, Italian Greyhound, Japanese Chin, Maltese, Toy Manchester Terrier, Miniature Pinscher, Papillon, Pekingese, Pomeranian, Toy Poodle, Pug, Shih Tzu, Silky Terrier, and Yorkshire Terrier.

In general, small dogs put out more heat per unit of body weight when they eat than larger breeds do. This diet manages the level of protein, since it gives off the most heat when consumed. Sodium bicarbonate (baking soda) is also included to help the smaller breeds tolerate heat better.

Another problem in the Toy Group, especially with Yorkshire Terriers, is hyperglycemia (high blood sugar). Chromium and niacin are added to help maintain blood sugar levels. Urinary tract infections are addressed in this formula by

Sugar and spice and everything nice. The Cairn Terrier is a member of the Terrier Group.

This Beagle member of the Hound Group knows how to carry his weight.

including cranberry extract to prevent e.coli and other bacteria from attaching to the urinary track.

Hound Group

The breeds included in this group are: the Afghan Hound, Basenji, Basset Hound, Beagle, Black and Tan Coonhound, Bloodhound, Borzoi, Dachshund, American Foxhound, English Foxhound, Greyhound, Harrier, Ibizan Hound, Irish Wolfhound, Norwegian Elkhound, Otterhound, Petit Basset Griffon Vendeen, Pharoah Hound, Rhodesian Ridgeback, Saluki, Scottish Deerhound, and Whippet.

This group is prone to intestinal problems and requires a highly digestible diet. Ground rice, rolled oats, and cracked pearled barley help provide a balance of digestible carbohydrates to improve gastrointestinal health. This diet also incorporates direct-fed microbials that are coupled with bromelain, a pineapple extract that provides enzymes that digest proteins and carbohydrates.

A 15 percent reduction from the standard feeding directions are recommended for breeds in this group like Dachshunds and Beagles to manage body weight and skeletal load.

NUTRITION AND SKIN, COAT, AND TEETH

In order for your dog to look good on the outside, you have to start with what goes inside. A high-quality diet containing the right amounts of proteins, fatty acids, vitamins, and minerals all help to provide healthy skin and a glistening coat. For the most part, problems from a nutritional deficiency are rare. If your dog's coat is dull, try switching to another good-quality food, but be patient. It takes one to two months before a change in diet—good or bad—will be reflected in the condition of his coat.

PROTEIN

A dog's hair and skin is made up of almost all protein. That's why it is important a healthy diet includes ample amounts of proteins like fish, lamb, and chicken. In order for the body to meet the demands of the skin and hair coat, adult dogs need a minimum daily amount of 18 percent protein, while puppies and mothers need at least 22 percent.

FATTY ACIDS

Besides playing a supporting role in providing a glossy coat, fatty acids also control inflammation and promote normal kidney function and reproduction.

A dog's body can produce some fatty acids on its own but the ones it can't—essential fatty acids—must be provided in his diet. In order to prevent deficiencies, at least two percent of the daily caloric intake should be from a fatty acid source, like linoleic acid. Feeding a good-quality commercial food will usually take care of this requirement. Owners should look for ingredients listed on

A dog's coat is often his pride and glory. The only way to ensure a glistening coat is to provide your dog with all the proper nutrients.

dry skin and dull coats. But for allergies and inflammation, supplements high in eicosapentaenoic (marine fish oil) and gamma-linolenic (evening primrose oil, borage oil, black currant seed oil) should be used. Approximately 20 percent of dogs with allergies benefit significantly with supplementation of these products as do some animals with degenerative joint disease (arthritis) and heart ailments.

Because of their anti-inflammatory effect, experts say fatty acid supplements can also help with other inflammatory diseases such as ulcerative colitis, bowel disease, and rheumatoid arthritis.

Pet supply stores carry supplements, or you can use a teaspoon of a vegetable oil like flax, soybean, safflower, or corn. But keep in mind if you give more than that, it can cause an upset stomach and diarrhea.

In general, dogs fed a high-quality commercial food will not need supplements because the proper balance of

This Irish Water Spaniel's handsome appearance is a match for the radiant flowers behind him.

the label such as fish oil, fish meal, or flax, which indicates the diet may contain a good balance of fatty acids. Deficiencies are uncommon in dogs, but when they do occur, owners will see signs of a dull, dry coat with scaling of the skin and hair loss. This sometimes occurs when dogs are fed poor-quality or low-fat food. A simple change in diet along with fatty acid supplements can help.

Keep in mind not all fatty acids work the same way. For example, supplements high in linolenic are recommended for

Vitamin A-responsive dermatosis is a skin condition that afflicts mostly Cocker Spaniels and can be treated with high doses of Vitamin A. This Cocker Spaniel is the picture of good health.

nutrients has already been added. On the other hand, dogs fed homemade diets probably will. Before adding any supplements to your dog's diet, consult your veterinarian first. Oversupplementation of one essential nutrient can adversely affect the absorption of others.

VITAMIN A

Besides playing a role in providing a healthy coat, Vitamin A affects growth, vision, and the immune and reproduction systems. A skin condition that afflicts mostly Cocker Spaniels—though any dog can get it—is called Vitamin A-responsive dermatosis. Signs include dandruff, hair loss, and marked crusting, especially on the back. This problem is treated by giving high doses of Vitamin A but only under a veterinarian's supervision, since it can be highly toxic.

VITAMIN E

Vitamin E is a natural antioxidant that protects fats in food and in the body from oxidation. It also acts as a mild anti-inflammatory agent and high doses have shown to help stimulate the immune system. Vitamin E is not toxic and a deficiency is unlikely as long as a good-quality diet is fed.

BIOTIN

This vitamin is used in the management of a variety of skin diseases, especially those that produce excessive scaling or dandruff. A biotin deficiency is rare, though it can occur when large amounts of uncooked eggs are fed. This is because raw eggs contain

Educating yourself on the proper vitamins and nutrients helps your dog live a longer, fuller life.

avidin, which prevents the body from naturally absorbing biotin. To prevent this from happening, simply cook eggs before feeding them to your dog.

ZINC

Zinc responsive dermatosis has developed in some Huskies and Malamutes due to their inability to absorb enough zinc from their diet. Signs include crusty skin sores and loss of hair around the snout and ears. Supplementation usually resolves this problem. Deficiencies also tend to occur when dogs are fed generic or high cereal diets. A change in diet helps solve this problem.

Here's a Tip!

The best time to evaluate the condition of a dog's coat is in the late spring when many have shed their winter coats and before others face a seasonal flea problem.

What signs should you look for? A healthy coat is shiny, feels silky, is of a normal thickness, and no excessive shedding or balding is apparent. An unhealthy coat is dull, feels oily, has bald spots, or shows signs of thinning in areas.

SUPPLEMENTING DIETS

Healthy dogs do not need additional vitamins and minerals when fed a complete and balanced high-quality commercial food. In fact, it can do more harm than good. Giving too much of one essential nutrient can adversely affect the absorption of others. Pet food is already formulated with the correct balance of amino acids, fatty acids, vitamins, and minerals your dog needs for a particular life stage. Haphazardly mixing in table scraps or supplements can upset this delicate balance. If owners want to increase palatability or add variety to their dog's diet, try mixing dry kibble with canned food or serving a combination of three different flavors of dry food. For owners who feed homemade diets, supplements like bonemeal, nutritional yeast, lecithin, kelp, vegetable oil, and several vitamins might be necessary. To find out what kind you'll need and how much to add, consult your veterinarian.

Besides homemade meals, supplements should also be given when there are signs of a nutritional deficiency, if organs are damaged, or if certain illnesses impair absorption of nutrients. Under these conditions, your veterinarian might advise supplements or recommend a special therapeutic diet.

TOOTH-FRIENDLY FOOD

What is considered a tooth-friendly diet varies depending on whom you talk to. Some experts say a dry crunchy kibble is best at promoting good oral care because the abrasion from chewing helps scrape tartar off teeth. Others say feeding one or two small chunks of raw meat daily is the answer. That's because tough cuts like stew beef or chuck steak contain large amounts of connective tissue that act like dental floss as well as

The kind of food you feed your dog affects the condition of their teeth. Some experts say a dry crunchy kibble is the answer, while others claim that feeding one or two small chunks of raw meat daily is the solution.

Two of a kind. This pretty pair of German Shepherds looks at ease taking a break.

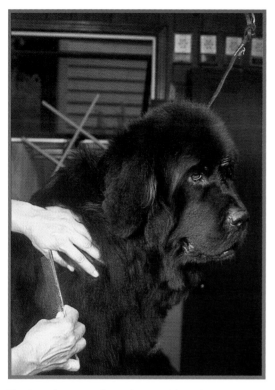

The best time to evaluate the condition of a dog's coat is in the late spring when many have shed their winter coats and before others face a seasonal flea problem. This Newfoundland doesn't seem to mind.

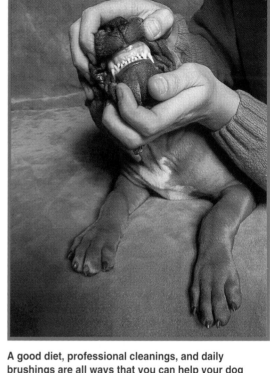

A good diet, professional cleanings, and daily brushings are all ways that you can help your dog achieve excellent oral health.

scrape away tartar. Still others claim that certain chemicals added to food are the best way to help reduce tartar build-up. One company adds sodium hexametaphosphte to some of its diets, which has been shown in clinical studies to reduce tartar accumulation by approximately 50 percent.

Early prevention through diet, dental checkups, and regular brushing are essential in helping your pet maintain good oral health, especially when you consider that more than half of all dogs begin to show signs of dental problems by the time they are three years of age. And smaller breeds, in particular Toy Poodles and Bichons, tend to suffer more from periodontal disease.

By taking early steps to get rid of plaque before it becomes tartar, periodontal disease can be prevented. A complete oral care program should include yearly dental checkups by your veterinarian. Some of the warning signs he'll look for are bad breath, a yellow-brown crust of tarter around the gum line, and pain or bleeding when the gums or mouth are touched.

If dental disease is found your veterinarian may recommend a professional cleaning to remove the accumulative plaque and tartar. At the same time, loose teeth will be removed and so will gum tissue if the pockets surrounding the teeth are loose or contain cauliflower-like growths.

Besides diet and professional cleanings, experts also recommend that you brush your pet's teeth daily. For most busy owners, this is not possible. A more realistic goal is once a week, but remember, only use toothpaste specifically formulated for dogs. It can be swallowed without causing an upset stomach like human brands.

By following a few simple steps good oral health can easily be achieved. And not only will that make your dog smile, but you as well—no more doggy breath!

NUTRITION AND HEALTH

If you thought dog food was just something to satisfy your pet's hunger, think again. Over the years, manufacturers have formulated special therapeutic diets that address health problems ranging from food allergies and obesity to more serious afflictions like renal failure. Through ongoing research at universities, as well as pet-food manufacturing facilities, nutrients known to control certain organ systems or metabolic pathways are added or reduced from food to lessen symptoms and make pets feel better. Most therapeutic diets are only sold through veterinarians. Listed below are some of the more common health problems that can be nutritionally managed.

ALLERGIES

Food allergies account for about 10 percent of all allergic reactions in dogs. But diagnosing it can be difficult since the symptoms resemble those of other canine allergies. The most common sign, though, is inflamed itchy skin (pruritus) around the feet, face, ears, armpits, and groin. Some will also experience loose bowel movement, diarrhea, and vomiting.

An allergic reaction to food can strike even if the same diet has been fed for years. It can happen at any age but puppies less than one year of age or adult animals older than seven years of age without prior pruritus are more likely to have an allergic reaction.

But before assuming the culprit is food, other possible allergies need to be looked at, like flea dermatitis, which is the most common cause of pruritus, and inhalant allergies, which are seasonal reactions to pollens and mold.

You can never be too careful when it comes to your pet. If something looks out of the ordinary, it's wise to take a trip to the veterinarian.

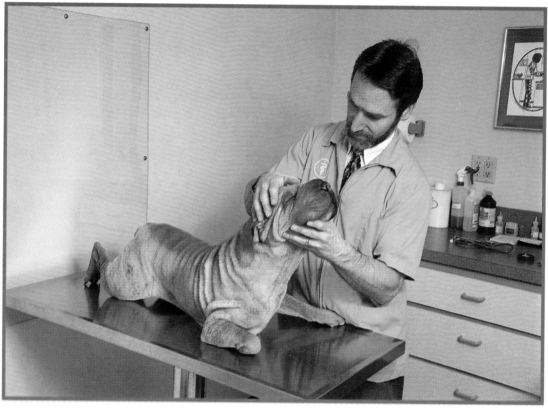

This Shar-Pei deserves a treat for his good behavior while on the examining table.

Once those are ruled out, a hypoallergenic diet should be fed to confirm that a food allergy exists. Most dogs are sensitive to only one or two ingredients, the most common being a protein source like beef, pork, chicken, or eggs. Therefore, experts suggest feeding a hypoallergenic diet made up of one protein and one carbohydrate the dog has never been exposed to. For years, lamb was touted as a protein source for this type of diet because many dogs had never been exposed to it, but today that's not the case. To fill the void, more exotic meats like rabbit, venison, or horse meat have been introduced as protein sources. For carbohydrates, potatoes, rice, or beans are commonly used. If you don't feel like making

your own, commercially prepared hypoallergenic diets can be purchased through a veterinarian.

Dermatologists recommend feeding the test diet for up to 10 weeks. During this time only the hypoallergenic diet and water should be given—nothing else—no treats, rawhides, or chewable medications. Remember to gradually introduce the test diet over several days by slowly mixing it in with decreasing proportions of the old food to prevent an upset stomach and diarrhea.

If your dog's condition improves during this test period, it's safe to assume one of the ingredients in his regular diet was the problem. But to be sure, some veterinarians recommend re-

introducing the original diet. Then if the problem starts up again within 7 to 14 days, it confirms a food allergy.

Once diagnosed, avoid all the ingredients in the old diet. If you want to know exactly which one is causing the problem you'll need to conduct the above test several times. First feed the hypoallergenic diet until the condition clears up, then re-introduce the old ingredients, one at a time, to see if the condition comes back.

Here's a Tip!

The University of California at Davis has a Nutritional Support Service staff that will answer questions from pet owners, clients of the veterinary medical teaching hospital, and referring veterinar-

It is a good idea to observe your dog while he is eating. Make sure he doesn't gulp or swallow large amounts of air to avoid experiencing a life-threatening problem called bloat.

the opposite. Nutritionists found feeding a diet with moderate protein levels does not have an adverse effect as previously thought. Moderate protein levels help animals maintain muscle mass, increase energy levels, and keep up normal activity.

Clinical trials indicated dogs in early stages can be fed about 50 percent more protein than found in conventional renal diets. And dogs in advanced stages can safely be fed approximately 25 percent more protein.

To feed these higher levels, fermentable fibers are added to food to divert nitrogenous wastes away from the kidneys and into the feces. Iams uses a unique dietary management tool called Nitrogen Trap & Trade and in an article about it, the company states, "Essentially it stocks the intes-

ians. They also have a computer program that develops and analyzes homemade diets for dogs. You can contact the Nutritional Support Service by calling the veterinary teaching hospital at 530-752-1393.

KIDNEY (RENAL) FAILURE

Renal failure is the inability of the kidneys to perform their normal function of filtering toxic waste products from the blood. For animals, chronic renal failure is a common illness that is irreversible and inevitably fatal. Carefully planned pharmaceutical and nutritional management can slow down progression and make patients more comfortable. For years renal failure was treated by restricting the amount of protein in the dog's diet, but recent research by the Iams Company shows just

It is important to provide your dog with a nutritionally fortified diet geared toward his stage of life. Look for foods that are naturally preserved, contain no by-products, and are 100% guaranteed. Photo courtesy of Midwestern Pet Foods, Inc.

This active Bull Terrier has the right idea. Increasing a dog's amount of exercise can reduce his risk of pancreatitis, a painful disorder that can strike any dog.

tines with bacteria that feed on urea. It works because fermentable fiber in the diet promotes proliferation of intestinal bacteria. These flourishing bacteria hydrolyze urea to ammonia and incorporate it into their own protein. When the bacteria are excreted in the feces, along with them go the nitrogenous wastes. The Nitrogen Trap & Trade works without sacrificing nutrient availability or stool quality. The unique blend of fermentable fibers allows feeding of higher protein levels without exacerbating uremia."

BLOAT

Dogs that swallow large amounts of air while gobbling down a meal or during heavy exercise can experience a life-threatening problem called bloat. Those most susceptible to this digestive condition are large and giant breeds like Great Danes, German Shepherds, Saint Bernards, Boxers, Weimaraners, and Standard Poodles. Bloat usually occurs after eating and causes the stomach to twist, trapping fluids and gas within it. Signs include retching without vomiting, excessive salvation, weakness, swelling of the abdomen, abnormal rapid breathing, and increased heart rate. Dogs experiencing any of these signs should be rushed immediately to a veterinarian since death can occur within hours. Each year 36,000 dogs are affected and of those, 30 percent die.

The exact cause of bloat is not fully understood but managing meals can help. If a dog normally eats dry food, moisten it with water. This way it will expand in the bowl instead of the stomach. Also, instead of feeding one or two big meals, serve several smaller ones throughout the day. For multi-dog households, feed separately to prevent them from eating too fast. Lastly, don't allow exercise an hour before or after a meal.

PANCREATITIS

This is a painful but often preventable disorder that can strike any dog, though overweight as well as middle-aged females are more prone. Pancreatitis is the inflammation of the pancreas, which is a small organ lying next to the lower part of the stomach, and is usually caused by consumption of a fatty meal. In mild cases, signs include lack of appetite, vomiting, and sluggishness. If you see these symptoms, experts recommend withholding food and

Best friends forever! Besides proper nutrition, love and affection also aid in a dog's overall health and well-being.

water for 12 hours so the pancreas has a chance to settle down. The biggest mistake owners make is to feed their dogs during this time, which only exacerbates the problem.

If signs subside after 12 hours, gradually re-introduce water and a small amount of your dog's regular food. But if the dog continues vomiting, suffers from shock (low blood pressure and cold skin), has difficulty breathing, or his heartbeat becomes irregular, call your veterinarian. In chronic cases, a diet moderately high in fiber may lessen the number or severity of attacks. Also having your dog lose some weight and increasing his amount of exercise can help.

DIARRHEA

Most dogs will experience soft or runny stool at least once during their lives. It can be brought on either by a sudden change in diet, by eating non-food items from those midnight garbage raids, or by gulping down too much of his normal ration. In most cases, it is resolved within one or two days without medical treatment. For these less serious cases, some veterinarians recommend skipping one or two meals to let the stomach settle down, but plenty of fresh clean drinking water should still be available. For the next meal, a bland diet is usually recommended for several days before re-introducing the regular fare. Remember when changing

diets to slowly add the new food in over several days. This will help reduce the chance of your dog experiencing an upset stomach or diarrhea again.

More serious cases of diarrhea are caused by parasites, toxins, or diseases. If your dog shows signs of a fever, experiences abdominal pain, vomits, seems depressed, or his stool contains mucus or blood, call your veterinarian right away.

CANCER

Cancer is an uncontrollable growth of cells within the body. Common signs are sores that do not heal, weight loss, an abnormal swelling that continues to grow, persistent lameness, and

Up to 40 percent of all dogs are considered overweight, making obesity the number one nutritional disorder. Make sure you allow your dog to get plenty of fun exercise, like swimming, as well as give him the proper diet.

difficulty breathing, urinating, or defecating.

Cancer is very common in older pets. In fact, it is the cause of death in almost 50 percent of those over 10 years of age. A few types of cancer, such as breast cancer, can be prevented by having your dog spayed or neutered between 6 and 12 months of age. Unfortunately though, most types cannot be prevented since their cause is unknown.

For over a decade, Colorado State University has conducted research in how nutrition plays a role in treating cancer. Recently, based on that research, Hill's Pet Nutrition came out with a cancer-fighting diet. The food is formulated to provide the nutritional requirements of dogs but at the same "starve" the cancer. This therapeutic diet is fed in conjunction with medical treatment, such as radiation or chemotherapy, and consists of lower carbohydrates but higher fats and good-quality protein. Researchers say tumor cells utilize carbohydrates for energy more than other sources such as fat; therefore, lowering the amounts of carbohydrates "starves" the tumor. According to researchers, when this particular diet is used in conjunction with a chemotherapy program, the survival rate skyrockets to 354 days compared to only 130 when regular food is fed.

Here's a Tip!

Some common types of cancers, like the ones listed below, can be treated if caught early. If you notice any of the signs—even though some of these are also seen in noncancerous conditions—

Ready, set, go! Frisbee™ is an activity that you and your best buddy can enjoy together.

consult your veterinarian right away.

Breast—Spaying between six and 12 months of age can greatly reduce the risk of breast cancer.

Head—Cancer of the mouth is common. Signs to watch for are a mass on the gums, bleeding, odor, or difficulty eating.

Testicles—For dogs with retained testes this type of cancer is common.

Abdomen—Signs to watch for are weight loss and enlargement of the stomach.

Bone—A common area cancer strikes are the leg bones, near joints. Signs to watch for are persistent lameness and swelling.

GAS

If this condition happens infrequently to your dog, two of the most common causes are eating table scraps or consuming large amounts of food in a short period of time. If this is the case, simply cut out the extra snacks or feed smaller, more frequent meals.

But if your dog is a chronic gas passer, take a good look at his diet. Food high in fiber, fat, or protein is a common culprit. Another cause can be from a recent change in diet. If this is the case, switch back to the old food for one week and see if it helps. If it doesn't, one expert recommends supplementing the diet with a plant-derived digestive enzyme or activated charcoal that absorbs toxins, gas, and other material that bother the intestines. Besides supplementation there are also therapeutic formulas your veterinarian can recommend.

OBESITY

Up to 40 percent of all dogs are considered overweight, making obesity the number-one nutritional disorder. It's not just the extra pounds that add up, it is also the health risks: locomotion aliments like hip and elbow dysplasia; arthritis; lower resistance to infections and viruses; liver disease; high blood pressure; not to mention skin problems and difficulties breathing. If that's not enough to convince you how serious the problem of obesity is, this should: Fat dogs die young. And since most owners would agree a dog's life is already much too short, don't let obesity rob what precious time you do

Flyball is one of the many fun exercises that your dog can participate in to ward off obesity.

have.

The main reasons why those extra pounds add up are from overeating and not enough exercise. Some breeds do have a genetic tendency to put on weight easier, like Labrador Retrievers, Beagles, Basset Hounds, Dachshunds, and Cocker Spaniels. Also, older dogs that are less active or a spayed or neutered dog can also suffer weight gain, not from the operation, but from the decreased activity level afterwards.

Dogs are considered overweight when they are 5 percent more than their ideal body weight and obese when they are more than 20 percent. But, just as with humans, not every dog is meant to weigh the same. Ideal weights should only be used as guidelines and each pet assessed on an individual basis.

To determine if your pooch is heavier than he should be, try this test: Place your

thumbs squarely in the middle of your pet's back and allow your fingers to feel the ribs. If you can easily feel them, with just a slight layer of muscle and fat in between, your pet is probably the correct weight. If the ribs are difficult to feel, with a thick layer of fat in between, your pet could be overweight. If you can't feel the ribs at all, your dog is probably obese. To be certain, though, visit your veterinarian to have him confirm that your pet is overweight.

If a few pounds do need to be shed, the first step is to change to a low-calorie food. Most veterinarians recommend a high-fiber formula because it has fewer calories and gives a "full" feeling, but some veterinarians disagree. They say a fiber diet reduces digestibility, increases stool volume, and decreases skin and coat condition because it doesn't provide enough fat and nutrients. Instead, they recommended a high-quality protein formula, so muscle loss does not occur, along with adjusted amounts of fatty acids to help maintain a glossy coat throughout the dieting process. Whichever food you choose, feed three or four smaller meals instead of just one so your dog will feel full and satisfied throughout the day.

Next, cut out the high-calorie snacks and treats. If treats are used as a reward, try replacing them with toys or praise. An exercise program should also be started. Take your dog for walks when the temperature is the most comfortable, avoiding times of day when it is extremely hot or cold. If your dog is not used to exercise, start off with two 15-minute walks per day and gradually increase the duration over several weeks. Remember, if there is excess panting or whining, you're pushing too hard. Another way your dog can burn some calories is to have him follow you around the house while doing chores. Every time you stop to do something like make the bed, tell him to sit. If he gets up, have him sit again. By doing a few of these "sit ups" each day it will help firm his hindquarters and abdomen.

Owners should keep in mind that a diet and exercise program requires a lot of time, effort, and above all else, patience. Results vary in each case and could take as long as four months before the desired weight is achieved.

Here's a Tip!

A diet tip for multi-dog households: If only one dog is on a diet, feed him separately from the others. This way you can be certain he is only eating the diet food and not "cheating" by munching on the higher calorie meal served to the others.

EXERCISE

Flyball, Frisbee™ and agility clubs around the country provide great ways for both you and your best friend to stay in shape and have fun while doing it. You might have caught a glimpse of one of these increasingly popular canine sports on television, but if not, here's a brief description about how each is played and how to get involved.

Agile and physically fit, "Lilly" soars through the air, exhibiting grace and strength.

The Weimaraner is part of the Sporting Group. This group usually needs extra calories and energy for exercise performance.

Agility

In this sport the handler directs a dog over a timed obstacle course. Mixed and purebred dogs of all ages are allowed to compete. They race against the clock as they jump hurdles, scale ramps, burst through tunnels, traverse a see-saw, and weave through a line of poles. Scoring is based on faults similar to equestrian show jumping. The United States Dog Agility Association (USDAA) has established four basic competitive classes based on jump heights. Dogs measuring over 16 inches in height must jump in the 24-inch or 30-inch class. For dogs measuring 16 inches or less in height there are two classes of competition—the 12 inch and 18 inch. Dogs measuring over 12 inches high but not more than 16 inches must jump in the 18-inch class. The different height classes were developed to provide safe but challenging jumps that are fair in competitions.

The best way to see if your dog will enjoy agility is to try it. Most dogs do well because the obstacles are relatively easy to learn and it doesn't require hours of training like some other competitive canine activities. To learn more about agility competitions or clubs in your area contact The United States Dog Agility Association at (972) 231-9700 or visit the web site at http://www.usdaa.com.

Flyball

Flyball was started in California in the late 1970s by Herbert Wagner, who first demonstrated it on the Johnny Carson show. Today the sport is widely played throughout the country. Flyball is a relay race with four dogs on a team. The game is played on a 51-foot course that consists of four hurdles—spaced 10 feet apart—and a spring-loaded box that shoots out a tennis ball. Each dog jumps the hurdles, steps on the box, and catches the tennis ball, then races back over the hurdles to the starting line where the next dog goes. The first team to have all four dogs run without errors wins the heat. Tournaments are usually double elimination or round robin format. Double elimination is best of three or best of five. Round robin is usually three out of five. To find out about clubs in your area or tournament information contact the North American Flyball Association at 1400 W. Devon Ave., #512, Chicago, Ill. 60660, or visit the web site at http://muskie.fishnet.com/~flyball/flyball.html.

Frisbee

It all started back in the mid-70s when Alex Stein crashed a Dodgers baseball game with his Frisbee™ dog "Ashely Whippet" and performed a high-flying demonstration in front of a nationwide audience. The crowd loved it and the sport of canine Frisbee™ was born.

Competitions like the Alpo Canine Frisbee™ Disc Championships are divided into beginner and intermediate levels with each consisting of two different events. The first is called the mini-distance, which is played on a 20-yard field and competitors are given 60 seconds to make as many throws and catches as possible.

The second is the freeflight event, which consists of a choreographed series of acrobatic moves to music. Judges award points on a 1 to 10 scale in each of the following categories: degree of difficulty, execution, leaping agility, and showmanship. Bonus points can be given to competitors with spectacular or innovative free flight moves.

Mixed or purebred dogs can compete but there are certain characteristics that help make a good canine Frisbee™ competitor like strong retrieval and tracking instincts, even temperament, a lean build and sound hips.

Friskies/ALPO sponsors over 100 community contests throughout the country each year. There are also seven regional qualifying tournaments culminating in the invitational World Finals at the mall in Washington, DC. To find out more about competitions in your area or for a free Frisbee™ training manual call 888-444-ALPO.

Playtime is fun! This little Shar-Pei and his friend are busy discovering the world.

NUTRITION AND HOUSEBREAKING AND TRAINING

With a little patience, your puppy can quickly learn to relieve himself consistently outside and not on your new carpet.

The first step is to set up a regular feeding schedule by serving meals at the same times and place every day. Up until four months of age, puppies will need four meals per day: morning, midday, mid afternoon, and an hour before bedtime. At four to six months of age, reduce to three meals a day: morning, noon and an hour before bedtime. From eight months on, feed once or twice a day. The amount you feed your dog is also important. Be consistent in the amount given. A handful or bowl full can mean different things to different family members who help with mealtimes. To feed a consistent amount, break out those measuring cups. Also keep in mind the serving portions listed on the label are for the entire day. To calculate a single feeding, divide the suggested amount by how many times a day (two or three) you feed your pet.

The quality of food also makes a difference. Higher caliber formulas are more dense in nutrients, which allows the body to utilize more and therefore less comes out. Avoid changing brands while housetraining because it can cause an upset stomach or diarrhea. If you must make a switch, do it gradually over several days by mixing increasing amounts of the new diet with decreasing proportions of

This trio of Bullmastiff puppies is positively irresistible.

Take your dog to the same place to eliminate when teaching him to go outdoors. He will soon know what's expected of him.

During housetraining be prepared to make a lot of trips outside. Puppies over 10 weeks of age might need to go outside a dozen or more times per day, while adult or elderly dogs require at least three or four outings. If at any point your dog whines or walks in circles while sniffing, take him outside immediately, since these are signs he has to relieve himself.

Once outside, always go to the same area. The odor from previous visits will help remind him that this is the place to go. Also allow him to sniff around for a while. This simple act helps stimulate elimination. Keep in mind that sometimes dogs will urinate or defecate more than once per outing, and not always right away, so be patient and don't rush him.

While outside it's important not to distract the puppy. Remember, he's there to do his business, not play. Once he eliminates always praise him by saying, "Good boy or good dog." The positive rein-forcement will help speed along the training process. If your dog does not eliminate, go back inside the house and wait 15 minutes, then take him outside to try again. Remember, until your dog is fully housebroken he needs constant supervision while inside the house. If you cannot watch him, confine him to a small room or area that has a hard floor. This way if an accident occurs, it will be easy to clean up.

If you catch your dog relieving himself inside the house say, "no" firmly, then immediately take him outside to finish. When he does, praise him warmly. If you get

the old until only the new is fed.

Meals should only be left down for 30 to 40 minutes then removed. Limiting meal times will make your dog's elimination schedule more predictable. Most tend to go an hour after eating, though really young pups might have to go after only 20 or 30 minutes. Also, dogs usually tend to relieve themselves after long peri-ods of play or naps, as well as first thing in the morning and just before bedtime at night.

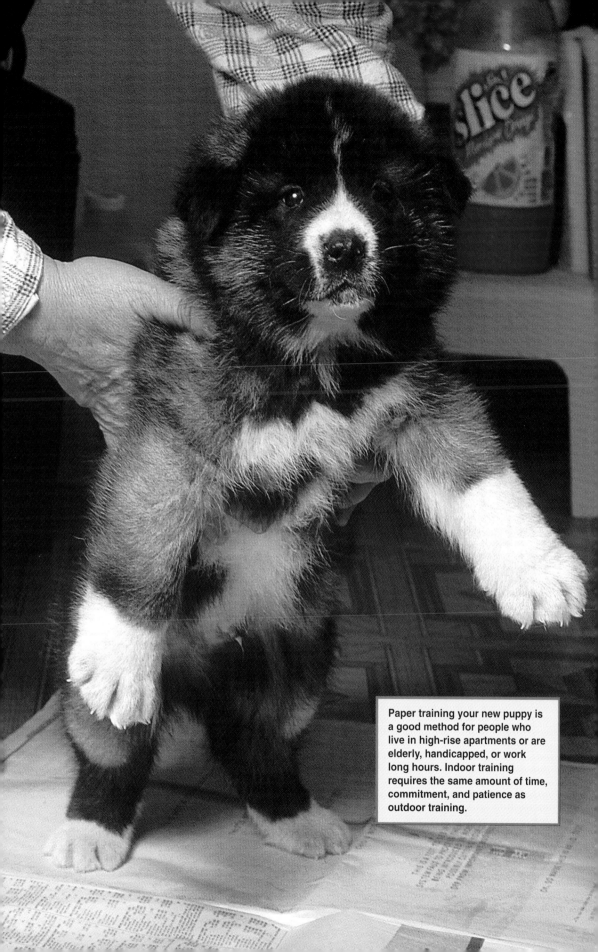

Paper training your new puppy is a good method for people who live in high-rise apartments or are elderly, handicapped, or work long hours. Indoor training requires the same amount of time, commitment, and patience as outdoor training.

"Down" is an important command that every dog should know. Remember that a dog that performs a trick correctly should always be rewarded!

your puppy for this because if you do, the problem will only get worse.

The key to successful housetraining starts with you. By sticking with a regular feeding schedule and making consistent and frequent trips outside, it can quickly be accomplished in a matter of weeks. If your puppy still makes mistakes after several weeks, consult your veterinarian. There could be an underlying medical condition such as internal parasites or a bladder infection hindering training. Also, don't expect to completely housetrain puppies under 14 weeks of age because they don't have full muscle control yet.

Housetraining Tips
• Stay on a regular feeding schedule.
• Only allow food to be available for 30 to 40 minutes, then remove it.
• Do not give table scraps or treats.
• Don't change diets; it could cause an upset stomach or diarrhea.
• Until housetrained, constant supervision indoors is necessary.

If you plan on keeping your dog outside, take the time to housetrain so you can still enjoy his company inside occasionally.

PAPER TRAINING
This method is good for people who live in high-rise apartments or are elderly, handicapped, or work long hours. Paper training allows your pet to relieve himself indoors on thick layers of newspaper instead of having to go outside. As far as indoor training goes, it requires the

there after the damage has been done, never rub his nose in the urine or feces. Dogs can only associate a reward or punishment with the act they are performing at the time, so he won't be able to understand that a mistake made 20 minutes ago is now making you upset.

Another important rule in training is to never hit your dog. Physical punishment accomplishes absolutely nothing and, in fact, only slows down the learning process. Also keep in mind urination due to submissiveness or excitement is totally involuntary. Don't discipline same time, commitment, patience, and regular feeding schedule that outdoor training does.

The first step in paper training is to decide on a location inside your house or apartment where your dog will be allowed to eliminate. Choose a spot away from his sleeping and eating areas. Once you establish an area, don't change it—this will only confuse your dog.

In the beginning, the selected area should encompass a three- to four-square-foot area (later this can be reduced) that is layered with newspaper or training pads. Take your puppy to this location to relieve himself after every meal, first thing in the morning, and last thing at night. Bring him there if he shows signs he has to eliminate, such as whining or sniffing while walking in circles. Over the next several weeks, with a lot of consistency and patience, your dog will learn exactly where to go without your help.

TRICKS AND TREATS

Owners love tricks and dogs love treats—who could have asked for a better match! And by initially using food as a training reward you'll be surprised how quickly your pooch learns.

After each trick is correctly performed, give your dog a treat by placing the piece of food in the palm of your hand, then present it in front of your dog's mouth so he can gently take it from you. Rewards should be small and easy to eat, such as tiny pieces of biscuit. Some manufacturers even make bite-sized morsels especially for training. Certain types of people food can also

Giving your dog a small treat for a behavior properly demonstrated reinforces the positive behavior.

work such as popcorn (hold the butter and salt), broccoli florets, baby carrots, green beans, apple slices, seedless grapes, blueberries, strawberries, and cooked pieces of lean meat.

Over time reduce the amount of treats given by allowing only one after every two or three tricks performed correctly. Eventually phase out food completely and just use praise as a reward. This is because too many extra calories can lead to obesity, which in turn causes health problems. Snacks, whether used for training or not, should only account for 10 percent of a dog's daily caloric intake. This doesn't mean you have to break out a calculator to keep track—just roughly figure it

out so you can reduce the calories consumed in snacks from his daily meal.

Listed below are several tricks to try your hand at, that is, as long as there is a treat in it!

Sit

This is one of the basic commands that all dogs should know and it is helpful when trying to teach new tricks. First, place a treat in your hand. Now place your hand in front of your dog's nose and slowly raise it above his head while giving the command, "sit." This will make him reach up to sniff your hand and cause his hindquarters to rest on the ground. When he sits, praise him and give the treat. Another way to

Rolling over is one of the oldest tricks in the book, but dogs always love to learn new things.

teach your dog is when you notice that he is going to sit, immediately give the command, "sit," then praise him and give the treat. Using this method, though, takes longer to train than the first one.

Down

Once your dog has mastered how to sit, the next basic command is "down." Start by having your dog sit. Then, with a treat in your hand, make a downward motion in front of his nose as you say "down." Only give him the treat if he does.

Crawl

Start by having your dog lie down. Then place your left hand an inch or two above his back. Your right hand (with a treat in it) should be about a foot away from the front of his face. Now tell your dog to "crawl" as you slowly move the food away. If he tries to stand up, gently push him back down with your left hand and repeat again. Once he crawls, even just a little, give him the treat and praise him lavishly. Practice several times a day and soon he'll catch on.

Find

This is a fun game with a lot of variations. Start by telling your dog to sit and stay. Show him one of his favorite toys, then place it somewhere in the same room where he can easily find it. Now ask him to get the toy by giving the command "find" or "toy." Once he goes to the toy, praise him. Keep repeating this, each time encouraging him to take the toy in his mouth. When he finally does, give him a treat and some praise. The next step is to get him to bring you the toy. Once he does, trade the toy for a treat.

After he's mastered that, the fun part begins. Have your dog stay in one room while you hide the toy or a piece of food somewhere else in the house, then ask him to find it. As he gets better, increase the level of difficulty by hiding the toy or food under things, behind different objects, or even outside in the yard.

Shake Hands

Start by having your dog sit. Say, "shake hands," then gently take his paw with one of your hands and give him a treat with the other. Practice several times a day over the next week. If at any point it seems like he is catching on, try giving the command "shake hands" and see if he'll give you his paw. If not, keep practicing.

Roll Over

Have your dog lie down on a soft surface such as carpet or grass. This is important since concrete or wood floors could hurt his back. Next, help your dog lie on one side. Then lift his legs and carefully turn him over while at the same time saying, "roll over." Only practice two or three times and make sure to reward with a treat and warm praise.

Training Do's and Don'ts
• Start training from 8 weeks of age on.
• Use short words for commands—sit, stay, down, come.
• Be consistent with commands. Make sure everyone in the family uses the same ones.
• Puppies have short attention spans so don't spend more than five minutes per training session.
• Master one command at a time then move on to the next.
• Be patient. It can take several weeks before your dog understands the commands.
• Don't yell, scream, or hit your dog. This doesn't accomplish anything. In fact, it can slow down the learning process.

Homemade Treats to Use in Training

The following recipes are from Dinger's Dog Bakery, the gourmutt bakery for dogs, based in Naples, Florida.

Yuppy Puppy Cookies:
1 small jar of baby food (any flavor)

Although these dogs look like they are searching in places where they are not supposed to be, the game of "Find" is fun for both you and your dog.

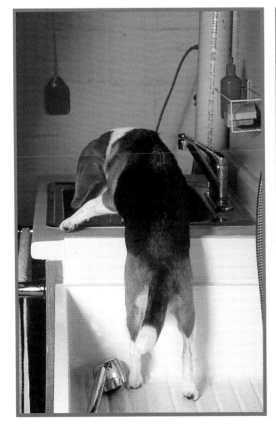

1/2 cup wheat germ

1/4 cup non-fat dry milk

1 teaspoon canola oil

Combine all ingredients in a bowl. If needed, add a little water. Roll into balls and place on a greased cookie sheet. Flatten with a fork dipped in non-fat dry milk. Bake at 350 degrees for 12 minutes. Store in an airtight container in the refrigerator. Makes approximately 12 cookies. Nutrition information per cookie: Calories 38, Fat 1 gram, Protein 3 grams, Carbohydrates 5 grams.

Bar-B-Que Bones:

2 cups whole wheat flour

$^1/_2$ cup wheat germ

2 tablespoons of brewers yeast with garlic

$^1/_4$ cup canola oil

$^3/_4$ cup beef stock

1 egg

1 tablespoon

Eating out of the garbage is a no-no! If this behavior occurs, put an end to it immediately. Simply place your garbage cans out of reach, and put the lids on gently.

Worcestershire sauce

Combine dry ingredients in a bowl. Add oil and beef stock and mix well. Roll dough out on a floured surface to $^1/_4$-inch thickness and cut with a bone shaped cookie cutter. Beat egg and Worcestershire sauce together and brush on cookies. Bake at 350 degrees on a greased cookie sheet for 30 minutes. Turn heat off and let dry in over for 3 hours. Store in an airtight container. Makes approximately 36 cookies. Nutrition information per cookie: Calories 46, Fat 2 grams, Protein 2 grams, Carbohydrates 6 grams.

PROBLEMS

Food Guarding

Some puppies learn early on that to eat they have to snap and growl at pushy littermates to keep them away from the food dish. As they become older this aggressiveness, if not corrected, can be directed toward family members. It can even expand beyond the food bowl to other things like a favorite toy or spot on the couch.

To solve this problem, start by having your dog sit. When he does, immediately give him a treat while at the same time saying, "take it." If he snaps too hard at the food, say, "gentle" in a firm voice and withhold the treat until he softly takes it, then praise him.

The next step is to repeat the exercise, but vary it just slightly. After your dog sits and you tell him to take the treat, wait a few seconds before giving it to him. If he

tries jumping on you to get the food, say, "no," and don't let him have it. By doing this your dog will quickly learn you have the right to control the treat. After he accepts this, the last step is to repeat the same exercise except this time offer the treat in his food bowl. Keep in mind this method should not be tried with dominant, older dogs that have a history of food-guarding behavior. In these cases, seek advice from an animal behaviorist.

Stealing Food

Your back was turned for just a second and now that sandwich you left on the counter is quickly being gulped down by your best friend. But it's not his fault. The savory smell of turkey and cheese wafting through the air is enough to make even the most well-behaved dog turn to a life of crime. And even though at first this may seem cute or funny, over time it won't be a laughing matter.

One way to solve the problem is to booby-trap a piece of food so that it will make a loud sound when your dog tries to take it. You can do this is by tying a thin piece of string to a soda can filled with several pebbles then attaching the other end to a hunk of meat. Next, strategically place the meat on the edge of the kitchen counter top. When fearless Fido gets one whiff and decides to go for it, he'll get a surprise. The noisy shaker can should give him a good enough scare to send him running out of the kitchen without the loot. If this doesn't work, try discouraging him by putting tabasco sauce, pepper flakes, or bitter apple on the food.

Eating Garbage

Trash cans are filled with spoiled, contaminated food and toxic chemicals that can cause your pet to become extremely sick, or in some cases, die. If your hound loves to hunt through the trash, place containers out of reach—it's that simple. In the kitchen, stash the trash under the counter. Outside in the yard, block off access to garbage cans with a portable fence (these can be purchased through pet supply stores and catalogues). And, as an extra precaution, make sure all trash cans have tight-fitting or locking lids.

Refusal to Eat

Lack of appetite could be the sign of a medical problem. Because you know your pet best, any deviation from his normal routine should be watched closely. If his lack of appetite continues to linger for several days, consult your veterinarian. But if it turns out your pooch is just a fussy eater, you'll need to make some changes.

One of the main reasons dogs develop selective taste buds is from frequent changes in their diet. Once you find a nutritionally complete and balanced kibble your dog likes, stick with it.

Feeding table scraps is another possible culprit. Not only does it fill him up so that he's not hungry, it also teaches him to hold out for something better than his regular fare. To entice Fido into eating his regular meal, try mixing dry and canned food together or moistening dry kibble with warm water or chicken broth.

Another possible reason for a lack of appetite is the weather. During the long hot summer months it's normal for appetites to slightly diminish. In fact, studies have shown dogs need 7.5 percent fewer calories with each 10-degree rise in temperature. So if it's mid-July, don't worry, your dog is just self-regulating his food intake.

Weather may have an effect on your dog's appetite, and it is not unusual for your dog's food intake to decrease in the spring and summer.

CONCLUSION

The adage "You are what you eat" holds true not only for us, but our best friends as well. There are six types of nutrients dogs need: protein, fat, carbohydrates, vitamins, minerals, and water. Providing the right amounts of each on a daily basis is essential for your dog's well-being, but figuring out how much to give doesn't have to be a mystery. Reputable dog food manufacturers have spent millions of dollars toward researching and formulating their prod-ucts to contain the correct proportions.

Great, you say, but how do I know which product to buy? With 15,000 different kinds of pet food competing for your attention, it can be challenging. Let's go over what we've discovered.

First, consider your dog's life stage: puppy, adult, nursing mother, or senior. As your dog goes through these different stages, his nutritional needs change, therefore so should his diet. Manufacturers understand this and have formulated specific diets to meet these changes. Before you buy a product, read the label to see for what life stage it is formulated.

On the label it should say "complete and balanced" or "complementary." Complete and balanced means it contains all the essential ingredients your dog needs on a daily basis to stay healthy and can be fed by itself as a meal. Complementary means that it is not intended to be used alone as

Knowing how to care for your dog properly is vital to his overall health and happiness. A good diet, exercise, love, and guidance can help your dog live a fuller, more satisfying life.

Bright-eyed and alert, this adorable Lab has a face that you can't help but love. Just be certain that before you take home your new friend you educate yourself on how to care for a puppy properly.

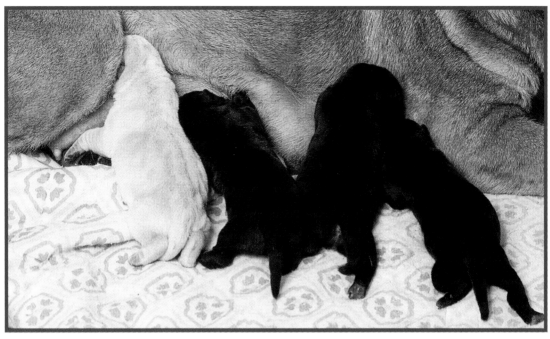

Puppies get the first nutrients they require from their mom. After they are weaned, it is up to you to provide them with a healthy diet.

a meal, it must be mixed with other food in order to create one.

Lastly, make sure the brand you buy is made by a reputable company. In general, manufacturers that have been around for awhile have had time to improve their product line, giving owners at least some assurance of a well-made product.

Now you've picked out a reputable manufacturer's brand that meets your dog's life stage and is complete and balanced. Great, but how do you know if you made the right choice? Keep in mind that there's more than one brand that can be right for your dog. There are two ways to tell if the product you've selected is working well in your dog's body. First, take a look at the stool. I know this isn't pleasant but two or three days after starting a new diet, take a look. You should see small,

firmly formed feces with no fluid leaking out.

Another sign to look for is a healthy, glossy coat but this will take some time before showing up. A change in diet—good or bad—takes one to two months before it is reflected in the condition of your dog's coat.

Instead of store-bought meals, many owners choose to make their own. Before running out to the grocery store, consult a reputable source for a nutritionally balanced recipe. Check with your veterinarian, or if you live near a university that has a veterinarian program, there's a good chance they have a nutritionist on staff.

Because of medical problems such as heart, kidney or cancer, some dogs must be fed therapeutic diets. These formulas contain nutrients known to control certain organ systems that are then added or

reduced from the food to lessen symptoms and make pets feel better. Most therapeutic diets are only sold through veterinarians.

Whether your dog is fed a commercial, homemade, or therapeutic diet, learning proper storage techniques are essential. Rancid food can cause vomiting, diarrhea, and other serious health problems. When storing dry food—opened or unopened—keep it in a cool, dry place. Canned products need to be refrigerated immediately after opening and used within two days. Look for a "best if used by" date on bags and cans as another means of ensuring freshness. More manufacturers, though not required, are now including these dates on labels.

There are two kinds of preservatives used in pet food: synthetic and natural. The most commonly used natural preservatives are Vitamin E and Vitamin C, but these are not as effective as synthetic preservatives and therefore shorten the shelf life of the product. Also, natural preservatives need to be used in larger quantities that, in the end, make a product more expensive.

Some common synthetic preservatives are BHA, BHT, and ethoxyquin. Over recent years, enthoxyquin has come under scrutiny. Some dog owners have claimed it causes a variety of problems ranging from allergic reactions to major organ failure. The Center for Veterinary Medicine has asked pet food manufacturers to voluntarily lower the maximum level of ethoxyquin used in dog foods from 150 parts per million to 75 parts per million. Studies are currently being conducted by the Pet Food Institute, a trade organization, to determine what is the minimum effective level of ethoxyquin in dog food. The study will be completed sometime next year. An official at the Pet Food Institute maintains that ethoxquin in pet food has never been proven to be harmful.

More than likely you'll change brands of food several times throughout your dog's life. When making a switch, slowly combine increasing amounts of the new diet with decreasing proportions of the old until only the new product is fed. This should be done over a seven to ten day period. This will prevent your dog from possibly having an upset stomach or diarrhea.

In general, a healthy adult dog can be fed just once a day. But how much to feed varies from dog to dog and from product to product. Owners should start off by following the feeding directions on the label, then increase or decrease as needed.

How much you feed your dog also plays an important role in housetraining. The first step in teaching your puppy to eliminate outside and not on your carpet is to establish a feeding schedule by serving a consistent amount of food at the same location and times each day. Meals should only be left down for 30 to 40 minutes, then removed. Restricting meal times makes your dog's elimination schedule more predictable. Most dogs tend to eliminate an hour after eating, though very young puppies might have to go after 20 minutes. Dogs usually have to relieve themselves after long periods of play, naps, first thing in the morning, and last thing at night. With a little patience and a lot of trips outside, your puppy will soon catch on.

Food is also a powerful motivator when teaching your dog new tricks. Treats given during training should be small and easy to eat. Some manufacturers even make bite-sized morsels just for training purposes and certain kinds of people food also work well.

Keep in mind that food does more than just satisfy your dog's hunger. It plays a role in every facet of his life—physically and mentally. It helps him fight off infection, maintain a shiny coat, keep up his energy level, and even helps to teach him how to behave. Love might not be spelled F-O-O-D, but it sure is close!

Providing your dog with the proper diet shows that you care about the quality of his life.

SUGGESTED READING

**TS-205
SUCCESSFUL DOG TRAINING**
Michael Kamer, OSB
160 PAGES, 130 FULL-COLOR PHOTOS

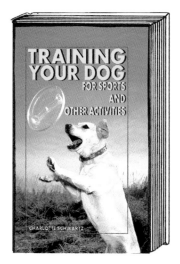

**TS-258
TRAINING YOUR DOG FOR
SPORTS AND OTHER ACTIVITIES**
Charlotte Schwartz
160 PAGES, 200 FULL-COLOR PHOTOS

**TS-249
SKIN AND COAT CARE
FOR YOUR DOG**
Dr. Lowell Ackerman, DVM
224 PAGES, 190 FULL-COLOR PHOTOS

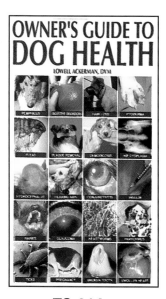

**TS-214
OWNER'S GUIDE TO DOG HEALTH**
Dr. Lowell Ackerman, DVM
432 PAGES, 300 FULL-COLOR PHOTOS